HYPERGEOMETRY

Introduction to Hyperspace Geometry

Manoel Carlos de Figueiredo Ferraz Parolari

ISBN 978-0-557-04693-5

www.parolari.eng.br

I dedicate this book
to my father and to my mother

I thank
John Dawla
for his help.

CONTENTS

Author's Comment

This little book was written in 1994. When I was a kid, I loved to chat about physics; my dad used to be my audience. My father died in 1985. I do remember quite well one of our last conversations about physics. I opined that time should have more than one dimension. Hence, the ideas championed in this paper have been mine for a very, very long time.

Hypergeometry used to be sort of a toy of mine. I took great pleasure in playing with my very own, exclusive toy. Quite a few men of science whom I had the privilege to know warned me that my attitude was a very selfish one and that I had no right whatsoever to play around with concepts that might appear to be very important to mankind, if not vital for the survival of our species.

This paper was written on this background, and in a climate of profound brokenheartedness for the loss of my beloved toy. I remained honest, though. Reading this book is an arduous undertaking, although it gives a substantial number of hints toward a better understanding of hyperspace. It was written for mathematicians, and it was written for the purpose of opening a dialogue on something I had discovered: how a child gives up his toy. It is sure not a textbook. I would like it to be a forerunner of a series of books written by brilliant minds who have mastered the art of teaching much more than I have. A small test, appended hereto, will show whether the reader has understood and has learned something. I have known for a long time that practical applications of hypergeometry are going to revolutionize human knowledge. There is no way to overemphasize the importance of the discovery of hyperspace. It further instructs a completely new way to behave with respect to our environment, our philosophy, and even our convictions.

I am convinced that this has been the missing piece toward comprehension of modern physics and that modern physics will bestow upon man the greatest possible expression of power: energy.

Developed nations foolishly hide such facts.
I have known for decades how important clean energy is to human development. The lack of clean energy will lead mankind to wars, famine, misery, and self-destruction. There is no lack of water or raw materials; we only lack large amounts of low-cost energy. Abundant low-cost energy will

satisfy all man's needs. The oceans may provide us with the metals we need. Their cost will depend on the cost of energy. Magnesium has been extracted from the sea for years, so any metal can be extracted.

Any metal?

Not really; plutonium is not to be found in nature. In my opinion, the name of this metal is very, very adequate and can only be the result of inspiration: the metal from Pluto, the Roman name that corresponds to Hades in Greek mythology, the lord of the kingdom of death. Right at the beginning of their research on radioactive materials, the Curies discovered that these bear the power and the seal of death.

I have been studying chemistry to a reasonable extent, decades of solitary study. When I was eleven, I had my own scholar chemistry lab as made by IBECC under UNESCO guidance in those days. If I had not had that very strong basis, which can only be obtained through endless experiments, I would not have been prepared to record Belight, the light of antimatter. Developing countries currently have great difficulties in getting hold of chemical reagents, as it is feared that such substances may be used to produce drugs and explosives. Merit is always in the middle, never at the outer ends. Such radical measures impede students to realize what chemistry is all about. The one who really masters chemistry will not have to give up by lack of the so-called forerunners. Students are the only ones to be disadvantaged; those who really master chemistry will make whatever they wants to make using substances that are impossible to be regulated. Anyone knows that cocaine can be refined with just cement and gasoline.

I am not referring to energy originating from radioactive substances. I do not accept nuclear waste that will not break down in less time than anything made by human beings. I am referring to clean energy based on modern physics. Governments are wasting huge amounts of money on research with low scientific value, while physicists like me do not have the most elementary equipment.

The consequences are obvious! Nations are despairingly in search of oil, that dangerous polluter, to satisfy their energy needs. Nations are fighting wars to keep up the geopolitical composition of forces and oil supplies. The great American nation has not been able to withstand the pressure of oil prices as high as US$147 a barrel. As a matter of fact, energy has triggered the 2008 financial breakdown of the developed countries, for they have not been able to generate low-cost, clean energy in due time.

There were no funds available to sponsor the right people, so these people could not do the necessary research work. Scientific institutions should learn to appoint individuals for their talent and competence.

We desperately need thorough knowledge of modern physics, for modern physics holds the key to extracting energy from matter.

In the meantime, this paper has been yellowing at the Library of Congress since 1995.

Brazil, January 1, 2009

Preface

There is no royal road to science, and only those who do not dread the fatiguing climb of its steep paths have a chance of gaining the luminous summits.

– Karl Marx, 1872

There are few places on earth where the above statement is more valid than in my country. One could in fact argue that Brazil has a rocky road through which science struggles to reach its heights. Amazing as it may seem, it is not for a lack of talented people, but primarily because of a nonsense view that science, although undeniably important, is too abstract to be used by practical people and should be left to the minds of those who have ability, passion, and time to *spare* with it.

This view developed not without cause. In a world where the basic needs of many have not yet been fulfilled and in a time when money is the objective of most, it is difficult to focus on issues that are apparently less accessible to the average human being. One would be surprised, however, with the real impact that even the most abstract concepts of science have in our daily life. Lord Keynes, writing about economists and philosophers, said that their ideas, "both when they are right and when they are wrong, are more powerful than is commonly understood". He continued saying that "practical men, who believe themselves to be quite exempt from any intellectual influences, are usually the slaves of some defunct economist". I believe it is not extrapolating too much to use his thought in an analogy to science in general.

In this context, with an evident inability to judge the contents of the book that I am so much honored to introduce, I take the opportunity to bring up other issues that are also very relevant to the occasion. It was not by coincidence that Marx was quoted in the opening statement of this preface – he who was in his own words the "best hated and most calumniated man of his time", despite the importance of his ideas to mankind. It is often the uncomprehended men who bring the most valuable contribution to science.

The author of this book has to a large extent also been an uncomprehended man. Confined in his farm in the Brazilian countryside, he endeavored in a lifetime of research, and he is finally able to share hits results with other scientists. In his road to science, he collected more sacrifice than profit. Completely isolated, he persevered with theories and ideas in the absence of both moral and material support. In his tireless efforts to develop his research, he often had to deal with intolerance, ignorance, and disdain. I hope this book will be the beginning of his intellectual harvest, rewarding him with the recognition that his intelligence and dedication deserve.

Ricardo Lacerda, New York, December 1995

Introduction

Inspire me, oh Muse, so that new paths to human knowledge and daring be discovered.

This is a little book that, for the first time in the history of geometry, does not present drawings. You may construct the possible drawings in your mind, depending on your capacity, as you read along. The new concept introduced is the possibility to do geometry *beyond the third dimension*! Multidimensional spaces have been known to man for a long time, but the prospects of geometrics beyond the third dimension have never been seen due to the simple limits of our physical space. Mathematically, however, multidimensional spaces *do* exist, as much as forms of understanding and executing typical geometrical operations, such as intersections, in more than three dimensions.

Euclid presents to mankind a geometry complete in itself: a plane geometry, or geometry in two dimensions. Little has been added to Euclid's geometry in terms of plane geometry. Centuries later, three-dimensional geometry raised into space what Euclid had written on planes. The fact is that a more ample generalization should have occurred, giving way to new postulates and axioms. But that generalization remained untouched; what *did* happen added little to the genius of Euclid.

Geo stands for earth, and *metron* for measurement; thus *geometron* is the "measurement of the earth" – in other words, the measurement of our houses, our fields, our backyards. As creatures that walk upon a surface, our ancestors essentially focused first on a geometry drawn on planes. Had we been birds or tree dwellers, we might have had the pleasure of discovering a three-dimensional geometry as our first geometry. Plane geometry is primarily the geometry of our bodies. Now, I ask you: what could man possibly draw better on paper than that which he drew on the very soil of his backyard, the venue of the birth of geometry?

Hypergeometron, where *hyper* means "above and beyond", identifies the geometry above and beyond the ground of your backyard. Indeed, it is the geometry born of the human mind, and as such, it can only be "drawn" within the human mind. The *topos* (the Greek word meaning "place") of our bodies is the ground, the surface. But the *topos* of the human spirit is infinity – infinity in the mathematical sense, in the sense of constantly and indefinitely showing us new possibilities. It is for this reason that I show no drawings, which can only exist mathematically or in the mind's eye.

Hypergeometry is not just a mathematical illation, but a tool that helps us to understand through more adequate models the world in which we live. As already mentioned, the *topos* of our body is our home; however, the *topos* of our mind is not. Human pride attempts to be present from the first moments of the Big Bang until the limits of the known universe. It tortures our intellectual spirit not to be able to look into the interior of an atomic nucleus. This does not deter us from venturing daring models that lead us to understand what happens in the innermost areas of matter that, despite being dimensionally very distant from our homes, can still be reached by our curiosity.

As a warming-up process, let us imagine, a two-dimensional ant living on an enormous sphere suspended in space. Let us analyse his limited world. He sees his world as one without frontiers, more so because he is able to measure its area. Based on domestic examples, his geometry will be essentially two dimensional. Now let us permit ourselves the right to play a trick on the little creature: let us place him, together with a companion, on a transparent Möbius band, made of a twisted tape whose ends have been glued together, thus giving it the quality of having only one side. Leaving his companion in one place, the ant sets out to explore his new environment, moving along both faces without crossing the edge. Imagine his surprise when he arrives at a certain spot and he sees his companion, still stationary, on the other face of the band. Within his two-dimensional world, the ant will never be able to explain the phenomenon of a two-dimensional tape with a three-dimensional twist! Only a geometry transcending beyond the concept of his domestic world (the surface of the sphere) could possibly enable him to understand what is happening.

Modern physics places an enigma before us that is similar to that of the ant when we observe phenomena in which time appears to move backward and particles escape from their seemingly absolute captivity. We are now faced with phenomena that can no longer be described through models produced with the knowledge that originated in our backyards.

Without a shadow of doubt, the progress of physics in the next millennium shall require hypergeometry for its new models in order to satisfy the insatiable curiosity of human nature. Should you be tempted to reject this new geometry for the simple reason that, for instance, you are not able to paint a four-dimensional space, do not forget that analytical geometry continues to be perfectly valid and that, as far as computers are concerned, the fourth dimension is a matter of just another index.

This work is the result of almost twenty-five years of dreams of a new path to physics that will once again uncover new horizons so far confined to a dead end – concepts that were, until this day, alive solely within my heart. This new geometry is an indispensable tool, as are the hammer and chisel to the sculptor. The aim is physics. It took less than a year to elaborate the written part. The book is little more than a child's game; in truth it is a series of new and interesting, but very simple, concepts. The ideas were developed to serve as the base of a much bigger accomplishment, a Herculean labour that will take more than a decade to write and whose role will be the application of the concepts to physics. Should this be accomplished, many other mathematical concepts that have already been created or extended but are isolated theorems will certainly emerge. The geometrical concepts presented herein contain a structure and form a body of knowledge that can be presented in a book. In fact, it is almost an appetizer.

I can hardly imagine how this piece of work shall be acknowledged. Like a jealous father, I am inclined to press it to my chest and protect it . . . and as a just father, I feel that I do not have the right to do so. Our children are not our property; they should be set free to face the world. The manner in which this work will be received shall pave the way to the elaboration and emergence of its sequel, which will definitely be a more toilsome process, requiring external help that was not necessary while writing this book.

Even as I write these lines at the heart of my little farm in the remotest Brazilian countryside, there are sounds of nature just outside my window: the twittering of birds, the rustling of green leaves caressed by the warm tropical breeze. In truth, my inspiration comes form the very peace that surrounds me, the liberty of a complete privacy where I am free to think and produce as I please. And it is precisely that liberty that I shall require to set forth the works that are meant to be of vital importance to mankind in the future. These works come from the heart, and they can only dwell in the heart while there is tranquillity around.

Those who receive this work with a welcoming gesture will certainly notice that the unfolding of its sequel shall require a support in the nature of finance – not finance to help its author live, but to help him create better conditions of work. This little book has been written during the intervals that come up between my work schedule as I coordinate the activities on my farm.

I can well imagine that the production of a massive work that lies stowed away in my mind shall require more time and constant engagement. As I require the quietness of my rural abode to produce, I am also dependent on it as an instrument of income. Were I to be partially relieved of my administrative activities, I should have more time and energy to dedicate mainly to the writing of books.

I hope to acquire the ideal conditions to work with all the power in my heart to bring forth a titan who may be able to contribute vitally to the development of a human knowledge and will one day bring abundance to our most precious wealth: energy. Though we are quite unconscious of it, we are surrounded by energy. In the act of extracting billions of tons of carbon from the breast of our planet, we pollute our biosphere for meagre amounts of energy, causing a greenhouse effect. So it is with our petroleum, with the coal in our earth and with the wood in our rain forests. Even more devastating is the case of our nuclear plants with their thousand-year radioactive waste, the product of an underdeveloped technology. Is it not time for man to invest generously in the development of a technology that would produce vast amounts of energy? Verily, this should be the great economic and strategic interest of the larger enterprises and of the governments of more developed countries. The very existence of mankind may depend on these investments, which may also yield gigantic financial returns. As an example, we may throw a quick glance to see the money generated within the industry of petroleum, which, beyond being just a raw material, is a vital source of energy. Were we able to produce energy from air and water, so abundant and so cheap, we could easily produce anything, from clothes to fertilizers, for this is really all that plants use – water, air, and solar energy.

Supposing we were already in possession of the technology required to produce energy at low cost and in great abundance, we would be joining water with energy to produce hydrogen gas, which in turn would produce steam in internal combustion engines. The burnt hydrogen would remove the oxygen released in the process of producing the hydrogen itself, and as a result, pollution would be absolutely zero!

All modern household appliances in developed countries consume energy. Inhabitants of colder regions know very well that heating is not just a matter of comfort but a condition to survival. Cost-effective and abundant energy can create better conditions of living for them, and a better standard of life for all nations. And the wealth generated by cheap and abundant energy will be fantastic!

I shall announce certain facts so that there may be no false expectations: I *do not* know a way to dominate energy of matter that may be used by man, although I *do* consider the new concepts expressed in my work a contributing factor to make it possible one day. I shall not expose these concepts in parts; they shall only be revealed in an integral publication containing all that has been developed in the past years and shall be developed in the coming years. It is through the experience of writing that we discover errors, and sometimes entire sets of erroneous equations, that should be rectified. Before releasing my complete work, I should like to assure that such corrections have been done and that the work is as free of errors as possible. I have given a name to the continuation of this work: *Physics of Matter.*

To conclude, I shall tell you a story, the story of how hypergeometry was born. When I was still a student, I became acquainted with Heisenberg's uncertainty principle, which may be translated as the following expression:

$$\Delta x \cdot \Delta p \geq h$$

where h is Plank's constant. If we wish to know and determine with precision the speed of a particle, we shall find ourselves in uncertainty as to its position. There are various interpretations to this uncertainty, one of which is based on Waves Theory, and considering De Broglie's wave-particle Duality, tries to accept this indeterminacy as natural. I have always been reluctant to accept these explanations. In any area of human knowledge, the normal procedure of science is to establish all the parameters that control a phenomenon, then vary just one parameter and observe the consequence. Varying another parameter, fixing it in a new condition, and returning to study the first parameter will show that there is no room for uncertainties, while there are enough parameters under control. If uncertainty h exists, it is probably because there are unestablished parameters. But which ones? A number of eminent physicists of our century have felt the need of other dimensions at the particle level, but no one has ever justified the reason for this hunch. And even more serious is that the mathematical tool to work with new dimensions of space did not exist. The missing tool was a new geometry. The missing tool is hypergeometry.

September 1995, Green Field Farm, Brazil

Chapter 1
Spaces: Symbols and Definitions

Axiom: a point is a space with zero dimensions.

A point P of an n dimensional space has zero dimensions.

Let us consider a single line. Let us imagine a space that limits itself to a single line. Everything that exists is within this line. In the line, let us consider a generic point that will be the origin: P_0. Having defined P_0, it is still necessary to define a direction and a measure; to do so, let us consider a second point different from P_0, hereafter called P_1. Given P_0 and P_1, we can refer to any other point of this space through the following parametric equation: $P = (P_1 - P_0)t + P_0$. The whole universe of one-dimensional points will be well defined through this expression. Note that the space in which we find ourselves can be of a dimension superior to one, and in it we can always consider an arbitrary straight line and restrict our universe to one dimension.

We propose for one dimension two distinct symbols that will prove to be equally useful. The one-dimensional space shall be named Alef (the first letter of the Hebrew alphabet) or E_1, referring strictly to one dimension. A point with zero dimension shall simply be called E_0.

To repeat what has already been said, using the proposed symbols: In a space E_n, we can always take a subspace Alef. It will simply be one when $n = 1$ or infinite when $n > 1$.

Definition:
Two spaces of dimension n are identical if and only if any point P belonging to one implies that it belongs to the other; in other words, all points of A belong to B, and vice versa.

Fundamental axiom:
Given two spaces A and B of dimension n: E_{nA} and E_{nB}. Given a subspace $E_{(n-1)}$ common to spaces A and B, if there is a P belonging to A and B that does not belong to $E_{(n-1)}$, then A and B are identical.

Theorem:
All space $Alef_1$ that has distinct points of $Alef_0$, is $Alef_0$ itself.

Proof:

It is sufficient to apply the former axiom. A subspace $E_{(n-1)}$ of E_1 is an E_0, or a point. So if two Alef spaces have a Point $E_{(n-1)}$ in common, a second distinct point in common makes the two spaces identical. Q. E. D.

In other words, if P_1 belongs to Alef$_A$ and Alef$_B$, and if P_2, which is different from P_1, also belongs to Alef$_A$ and Alef$_B$, then the spaces Alef$_A$ and Alef$_B$ coincide and have infinite points in common.

Note again that we can be in a space E_n in which we consider an initial point of origin P_0 and then any second point of E_n P_1. These two points define a one-dimensional space Alef, where a generic point of Alef is given by $P = (P_1 - P_0)t + P_0$ and where t is a real number .

In an n dimensional space (greater or equal to two), let us take a straight line r and a point P that does not belong to r. P and r define one and only one Alpha plane. An Alpha plane is a two-dimensional space. We are in the backyard of our house. Any other Beta plane defined by the same straight line r and the same point P, shall be identical to the Alpha plane according to the fundamental axiom.

Using the proposed symbols, a plane is a two-dimensional space E_2, hereafter also called Bhet space. A Bhet space contains an infinite number of straight lines, or Alef spaces. Within Bhet, let us take a straight line r. Any other straight line of Bhet may contain one or no common point with r, because if it contains more than one it will be identical to r.

Definition:
Two straight lines are crossing lines if they contain one and only one common point.

In the case of Bhet space, two lines meet only if they are not parallel to each other. Observe that we are generalizing the classic Euclidian geometry without considering infinity as a meeting point of parallels. We shall study the notions of crossing spaces and parallelism in chapter three.

Departing from the plane and taking a single point P that does not belong to the plane, we shall obtain a three dimensional space: E_3 also called Guimel space. Within a Guimel space there are an infinity of Alef and Bhet spaces.

To take one more step in the same direction of thought, we shall require a single point that does not belong to our physical space in order to jump to the fourth dimension. In the world we live in, this point does not exist, and the spaces stop right there. But you can imagine a point outside of our physical space that may even be quite near us and yet have *one* different dimension. From the mathematical point of view, there is no problem whatsoever. Taking

Point P(0,0,0,1), it cannot be existent in the space known to us, and that is exactly where the problem lies! This point exists in the mathematical point of view, though, and it can be easily expressed. This point P, which we have just considered, opens up a new four-dimensional space: E_4, or Daled space. In the same way, within a Daled space there is an infinity of Guimel spaces. This transition is the hardest because from here on a new universe is introduced to us: E_5 or Hei space, E_6 or Vav space, and so on into infinity . . .

Orthotropic Spaces

We should define here a concept that is simply an extension of Euclid's idea of points, straight lines, and planes. Euclid said, "A plane surface is a surface which lies evenly with the straight lines on itself."

Our spaces should always be expressed by a system of axes formed by reciprocally orthogonal lines that should not be limited.

Concept:

A set of all the points defined by a system of unlimited, mutually orthogonal axes may be called orthotropic space. A system of Cartesian axes defines an orthotropic space. This does not mean that an orthotropic space cannot be expressed by a system of cylindrical or spherical coordinates. What really matters is that it can be expressed by unlimited orthogonal coordinates.

I will give some examples. The open interval $(0,1)$ is an Alef space with infinite points. We may even imagine a function that maps points of this interval with an infinite straight line; therefore this space is not an orthotropic space because it is limited and because there is an M point larger than any point of the interval.

Studying the example of three-dimensional space defined by spherical coordinates, we shall see that it is an orthotropic space because it is unlimited and can be expressed by Cartesian coordinates.

Another example is a two-dimensional space defined by a circle without a circumference. As in the already presented case of the interval $(0,1)$, we can find a function that puts each point of the plane in correspondence with the points of this space that is not orthotropic because it is limited. We can find an M point in the X axis, larger than the x coordinate of any point of this space.

Looking at yet another example, this time an unlimited space, let us consider the surface of a paraboloid of revolution. You will never manage to lay two orthogonal axes to directly map the points of this surface.

To sum it up, the orthotropic spaces known to us are the point, the straight line, and the plane. We can, of course, imagine a multitude of plane deformations that are unlimited two-dimensional spaces, but they are not orthotropic spaces. Our physical space, as we know it, is an orthotropic space. Could there be non-orthotropic, three-dimensional spaces in existence? The answer is yes. When at a speed nearing the speed of light, the physical space that we know suffers deformations and ceases to be an orthotropic space. So what are orthotropic spaces really good for? The truth is that they are primitive concepts on which we can construct any other structure in any dimension.

Let us now establish a few concepts and dismiss natural doubts that emerge when we examine for the first time spaces with more dimensions than the physical space we are familiar with.

The first doubt: What, for example, is a plane in a Daled space like? Would it perhaps be a hyperplane? The answer is no. Let us not get carried away by the number of coordinates. A point is and shall always be a space of zero dimension, no matter where it is placed, the same way a straight line is an Alef space wherever it finds itself. And how does analytical geometry express a straight line? Answer: $P = (P_1 - P_0)t + P_0$ where P simply has coordinates $P(x_1, x_2, x_3, \ldots, x_n)$. Let us not be misled: a straight line cannot change its own properties through the acquisition of strange qualities; a straight line is always straight and always a line! Hyperlines do not exist.

All of Euclid's geometry remains valid. For example, two straight lines crossing one another define a plane – a straight, infinite (orthotropic) plane as we know one. Would you like to see it?

Let us call upon the meeting point of the lines r and s of P. No point of s, except P, can belong to r because there is a theorem that claims that all $Alef_A$ space that contains two distinct points of $Alef_B$ is $Alef_B$ itself. So there should be a point Q belonging to s that does not belong to r. We know that an E_1 (Alef) space and a Q point not belonging to E_1 defines an E_2, a Bhet space that is a plane. Q. E. D.

In this plane, as has already been pointed out, there are straight lines as we know them, and this plane can in fact be expressed by a sheet of paper or even the ground in your backyard, thus maintaining all of its geometrical properties. Since a plane is a Bhet space and a Bhet space is orthotropic, we can find in this Bhet space at least one pair of orthogonal axes X and Y and, based on these axes, apply everything we know about analytical geometry, writing the equations of curves known to us in the most familiar manner possible. And what about the crossing of other spaces on this plane? The crossings of orthotropic spaces are also orthotropic spaces, and the orthotropic spaces contained on the plane are points and straight lines. These are the only possibilities we shall ever find, as can be seen in chapter six.

Although it may seem strange when, for the first time, we are faced with multidimensional spaces, our daily life shows us at every moment functions that cannot be expressed in our three-dimensional space. Let us consider a tree. The function that describes the surface of a tree is a function of three variables: x, y, and z. The tree can be expressed in our space. And what if we wanted to mathematically express the colour of the surface of the tree? We could use the art of splitting the colours of each point into three numbers defining the intensity of magenta, cyan, and blue. We would then have a total of three more coordinates, which would result in the colour of the surface of the tree. It is of course a trick, but what counts is that we have managed to express the surface of the tree with its position and colours. But the tree is not immutable in time. If we looked at it once a month, we would observe in it the segment of time that represents a month. In reality, we would be using a seventh temporal dimension to express the tree. From a mathematical point of view, there should be no difficulty whatsoever to express a solid body, such as a sphere, in four dimensions. Neither should there be any problem to express it intellectually, once we are capable of imagining that this sphere varies with the passage of time. We cannot express this sphere with these temporal variations in a Guimel space. Well, what about it? As we shall see, our daily experiences take us to the principles of multidimensional spaces. But it is important to systematize, conceive, and appreciate in abstract form what is possible and what is not. And that is precisely what this new geometry is all about.

Comments:
Let us make a series of observations to familiarize the reader with the new mathematical concepts.

In this chapter, the fundamental concept is the concept of orthotropic space. Euclid defined points, straight lines, plains, and surfaces. Mathematicians that followed preferred not to define points, straight lines, and planes, claiming that these were basic and undefinable concepts. I quite appreciate the difficulty in defining them, but I would deem it more positive to try to describe their essence rather than do nothing about them. If someone should know of a more perfect description of the essence of points, straight lines, and planes, he should expose the same and extend it to other orthotropic spaces. According to Euclid, "A plane surface is a surface which lies evenly with the straight lines on itself." On the plane surface, the straight line, fixed to a point, can turn in any direction, being that all its points actually belong to the plane surface. This method can be formally recognized from the mathematical point of view and prove to be of great use. On the plane surface, let us consider a first point P_0. From P_0 and any other second point P belonging to the plane, the straight line defined by P_0 and P will be on the plane, and any point of this straight line will belong to the plane.

This concept can be generalized with Guimel space. From a generic point P_0, any other point P belonging to this space shall define a straight line completely contained in the space.

In Greek, *orthos* means "straight". Orthotropic space is defined by *straight* lines.

Definition:
Given a space E, within the space E let us consider a generic point P_0. E would be an orthotropic space if for any other point P belonging to E, the straight line defined by P_0 and P were completely contained in space E. And in order to maintain the generalization, E_0 shall, by way of definition, be called an orthotropic space.

Note that in Guimel space, the subspace of $n - 1$ dimension is the plane. Considering Daled space, the subspace of $n - 1$ dimension is Guimel space. The Guimel space is an orthotropic space like the plane, which is a new geometrical entity! Euclid referred to point, straight line, and plane; from now on we can refer to E_0 and spaces Alef, Bhet, Guimel, Daled, Hei, Vav . . . all containing properties that the same point, straight line, and plane have led us to recognize.

When you were told of a new geometry, you must have, from the first moment, imagined new geometrical entities. You were right; you will see geometrical entities that we shall call structures, something beyond solids (figures, solids, and structures), but the new and most important concept is that of orthotropic space. You may think you know Guimel space because we live in it, but the concept of Guimel space as a geometric entity is completely unheard of. You can hardly be too shocked by this new concept.

In geometry, you may have known crossing planes. The planes were well-known geometrical entities. In chapter three I shall introduce to you the crossing of two spaces, E_3. These crossings can be a simple point, a line, or a plane! And the crossing of spaces of superior dimensions may be a Guimel space!

This chapter is meant to familiarize you with multidimensional spaces. But how far can the number of dimensions rise? In truth, there is no limit to the number of dimensions, and this fact leads us to a new and equally important concept. The Greek word *cosmos* means "everything". We use the word "universe" with a similar meaning but without the magnitude of the word "cosmos". To give an example, the order does not belong to the universe, but in its true sense as the ancient Greeks understood it, the order is held within the cosmos.

From the mathematical point of view, a point P with an infinite number of coordinates is a cosmic point.

Definition:
Cosmos is the set of all possible cosmic points.

Thinking about multidimensional spaces, I imagine a part of the cosmos. Besides being tremendously abstract, the idea of cosmos is hard to conceive, but it does help to place us in multidimensional spaces.

There is a mathematical concept related to the magnitude of infinity. The number of real numbers is called $\aleph_0$ (Alef Zero). $\aleph_0$ is also the number of points in a straight line. The number $\aleph_0$ has some properties. For example, $\aleph_0 + \aleph_0 = \aleph_0$. Any real number r times Alef Zero is $r \cdot \aleph_0 = \aleph_0$. The only way to change the result $\aleph_0$ is by multiplying $\aleph_0 \cdot \aleph_0 = \aleph_1$. Some mathematicians consider planes as containing $\aleph_0$ points. That is not the way I see it. For each point of the axis X we have ($\aleph_0$) points in the axis Y, and therefore we have here a product of type $\aleph_0 \cdot \aleph_0 = \aleph_1$ points. In the same way, E_3 has $\aleph_3$ points. So how many points do we have in the cosmos? The answer is: $\aleph_\infty$ points. As we can see, the concept of cosmos blows up the concept of infinity. But the greatness of cosmos goes much further.

Given two cosmic points with varied coordinates, or at least with a *finite* number of identical coordinates, the distance between these two points will always be infinite. Although we shall study the generalized definition of distance in the next chapter, we may already perceive that the distance between two cosmic points whose coordinates differ from even the smallest value will be an infinite series of positive terms and therefore tend toward infinity.

All points are cosmic; however we shall use them without their cosmic qualities. The cosmos serves as a limit condition. A point with a very large number of coordinates is not a complete cosmic point. For it to be a complete cosmic point, it must be considered with an infinite number of coordinates. Any point of our dear Guimel space belongs to a complete cosmic point, just as any point of our backyard (a two-dimensional space) belongs to Guimel space. In the same way, if you draw a straight line on the ground and establish an origin, a direction, and a measurement, each point of the same straight line can be expressed by *only* one real number, even though the abstraction of the other coordinates is done from this point and is considered within the Guimel space. When considered within a space of limited dimensions, a point does not lose its cosmic properties!

All known points are cosmic points. Consider the point located on the tip of your ballpoint pen. It is obviously a cosmic point, or contained within a complete cosmic point. Consider a hypercube containing one micron of length of one side, based on the point defined by the tip of your pen.

What is a hypercube? A square (a two-dimensional cube) is expressed by all points $P(x,y)$ satisfying $x_0 < x < x_0 + \Delta$ and $y_0 < y < y_0 + \Delta$, where (x_0,y_0) is the coordinates of the initial point and Δ is the side of the square. In this way we can also define a traditional cube in a Guimel space based on the $P(x_0,y_0,z_0)$ as a set of all points satisfying:

$$x_0 < x < x_0 + \Delta$$
$$y_0 < y < y_0 + \Delta$$
$$z_0 < z < z_0 + \Delta$$

where Δ is the length of the side of the cube.

It becomes easy to generalize the process for any hypercube (square, cube, hypercube), where Δ is the length of the side, as a set of points that satisfy a given condition. This condition is to be placed between the given point and a certain increment Δ equal for all coordinates.

Let us return to the point at the tip of your pen where there is a cosmic point P_0. Now let us consider a cosmic hypercube whose length of side is one micron. You will easily map any little cube of one micron of your pen to a three-dimensional cube contained in the hypercube of one micron located on the tip of your pen. A hypercube possesses an infinite number of edges, and every three edges define a three-dimensional cube. In this way, there is an infinity of three-dimensional cubes on the very small hypercube located on the tip of your pen. What we have just done has been to put the whole pen inside the little hypercube, and yet there is a great amount of room left. You can even accommodate your whole room inside the little hypercube. To go a little further, why not put the earth and also the sun inside . . .? You could actually go on to put all the planets inside, and yet there would be infinite room remaining. So you can actually map any cube of one micron of our solar system to one cube of this hypercube. Go ahead and put in our galaxy or the whole known universe, and we shall still have a lot of infinite space left . . . Yes, the whole known universe fits in a small cosmic hypercube, no matter how small its edges.

In all probabilities, the first idea to emerge in your mind is the Big Bang. You may be looking at a mathematical form of how our universe may fit in a place that, by our dimensions, can be termed very small. The Big Bang is well beyond that . . .

The purpose of this reasoning is to introduce to you dimensions more ample than the world we live in. We are submerged in the cosmos; all the points of the world we know belong to more ample dimensional systems, whose geometry is only beginning to be known.

Before closing the chapter, it may be relevant to make one important observation. It was said in the first axiom that a point is a space of zero dimension. Later, a point P in E_n was expressed through n real numbers. Isn't there a conflict here? The answer is no. A point of zero dimension can, in turn, be situated in another space, and the location of this point shall demand as many coordinates as the number of dimensions of the space where the point is situated. Do not mistake space dimensions for the number of coordinates. For the sake of clarity, let us take our reasoning to an extreme. No coordinate shall be required to express whatever is contained within a space of zero dimension – E_0. Given an E_0, let us suppose that we could place a point there; how many coordinates would be required to express the location of this point? Not one; zero! And that is precisely what zero dimension means. But a point located in a Daled space shall require four coordinates to be expressed. The point has no dimensions at all, while the Daled space has four. Allow me to give you a little piece of advice: never use a space with more dimensions than you need. Just because you will now start working with n dimensions does not mean that you need to place yourself in n dimensions or even in the cosmos. The more dimensions the space in your study contains, the more coordinates your points will have. Your study will become unnecessarily complex, making you more vulnerable to mistakes. Do just the contrary: simplify as much as possible. Work with the fewest possible dimensions. You will only make it a lot easier on yourself.

Chapter 2
A Little Analytical Geometry

This chapter does not present innovations; indeed, it deals with the generalizations badly required after so many new concepts. Certain concepts should be redefined, and we shall demonstrate, very quickly, that analytical geometry remains essentially the same. In one of the chapters that follow, we shall show multidimensional structures that were not possible in Guimel spaces; however, the limit now is infinity. Everything presented in this chapter is quite general and is valid for any dimensions, but on many occasions I shall use Hei spaces, where points are expressed by (m,n,o,p,q) with m, n, o, p, and q are real numbers.

We have already seen that a straight line is defined by two points: an E_0 space (that is already a point) and one more point. Let us name these points P_1 and P_2, and let us show the mathematical expression of a generic point P belonging to the straight line. The general form of P will then be:

$$P = (P_2 - P_1)t + P_1$$

where t is a real number.

Example:
Let us suppose that P is given by coordinates $x_1, x_2, \ldots, x_n$; P_1 is given by $a_1, a_2, \ldots, a_n$; and P_2 by $b_1, b_2, \ldots, b_n$. In order to express the straight line, we must use a set of equations:

$$x_1 = (b_1 - a_1)t + a_1$$
$$x_2 = (b_2 - a_2)t + a_2$$
$$\ldots$$
$$x_n = (b_n - a_n)t + a_n$$

Observe that only in a Bhet space can we eliminate t and obtain the well-known equation $y = ax + b$ for the line. This does not mean that the line in a multidimensional space differs from the Euclidian straight line. The line is always straight and always a line; to refer to a hyperline would be folly, for this does not exist! All straight lines are equal to the lines we know, even though in a general case you could not make a drawing of it. The straight line is always an Alef space, defined by two points that do not coincide.

It is customary to define the distance between two points. We can go on to the generalization of the concept of distance between two points P_1 and P_2 without any hindrance:

$$d(P_1, P_2) = \sqrt{\sum_i (b_i - a_i)^2}$$

We can see that this expression is well-known. It always gives us a positive real number for two distinct points located in a space of any dimension.

To generalize, if P is a generic point of a hypergeometric structure A, and Q is a generic point of a hypergeometric structure B, then the distance from A to B is the minimum possible of d(P,Q).

Based on the definition of distance, we are able to define the magnitude of vector. But we should put that off until we reach the next chapter, which deals with vectors. With the exception of a few adjustments, analytical geometry continues to be the same as it has always been; even the mathematical forms of its expression have changed very little.

We already know how to jump out of a straight line and onto a plane: all it takes is to consider a point that does not belong to the line. The line, together with this new point, will define a Bhet space, and a Bhet space is no more than our good old friend the plane, the ground in our backyard. Here too, you should observe the warning related to the line: there is no such thing as a hyperplane; any plane is expressed by three points that do not coincide and are not collinear; in other words, the plane is defined by a straight line and a point.

Analytically, if we take a straight line by two points P_1 and P_2, based on only one of them – P_1 for example – and another point P_3 belonging to the considered space, we shall be able to draw another line defined by (P_1, P_3) and then have two crossing lines meeting at P_1. These two lines will define a plane that we shall call α. The α plane (a Bhet space) is a plane like the ground of your backyard, and you know how to work with it. Nothing can stop you from reasoning on the α plane, as we shall see in chapter six. Apply on the α plane everything you know about Euclidian geometry. As has been repeatedly said, hypergeometry is the generalization of Euclidian geometry. On a general level you cannot use particular forms, such as $y = ax + b$. But reasoning on the XY plane, you will once again be able to use exactly the same equation $y = ax + b$y=ax+b.

The analytical expression for the plane defined by P_1, P_2, and P_3 is:

$$P = (P_2 - P_1)r + (P_3 - P_1)s + P_1$$

On the above mentioned α plane, we can draw a third straight line based on P_3 and a generic point P belonging to the line defined (P_1, P_2). We can now look for the condition of parallelism or the condition of orthogonality. The equations shall lead to laborious systems, but there are computers to do the job for you.

As we insist that the α plane is our old acquaintance, we would like to work on it a little, just to remind ourselves of how old concepts continue to be valid. Once again, beware; we are in n dimensions.

The expression of parallel lines can generally be said to teach us Analytical Geometry. Given a straight line defined by P_1 and P_2, another straight line defined by Q_1 and Q_2 will be parallel to the former if

$$Q_2 - Q_1 = (P_2 - P_1)t$$

We know that P_1, P_2, and Q_1 define a Bhet space that we can call α plane. If our concepts of Euclidian geometry stay unchanged, even though these points belong to a multidimensional space, two parallel lines define a plane, and the generic point Q_2 belonging to the second line would necessarily be on α.

The expression of the α plane defined by P_1, P_2, and Q_1 is:

$$P = (P_2 - P_1)r + (Q_1 - P_1)s + P_1$$

making $s = 1$, we shall obtain:

$$P = (P_2 - P_1)r + Q_1 - P_1 + P_1 = (P_2 - P_1)r + Q_1$$

the condition of parallelism imposes $(Q_2 - Q_1) = (P_2 - P_1)t$

therefore
$$P = (Q_2 - Q_1)u + Q_1$$

which is exactly the expression of Q_2.

Once again, remember that we are in n dimensional space and that $P(x_1, x_2, \ldots, x_n)$ is expressed by a set of coordinates. In spite of this, on α plane things occur as we expect them to.

Now let us discuss parallelism for a while. Let us take an α plane defined by three points, P_1, P_2, and P_3. From a point Q_1 of the Space, so that Q_1 does not belong to the α plane, let us draw a straight line r so that r is parallel to $\{P_1,P_2\}$. Then the expression of r will be:

$$Q - Q_1 = (P_2 - P_1)r$$

In this line r, let us consider Q_2 so that $Q_2 \neq Q_1$. Thus, due to the conditions of parallelism, we obtain $(Q_2 - Q_1) = (P_2 - P_1)r$. In exactly the same way, let us consider a point Q_3 belonging to a line s passing through Q_1 that is parallel to a straight line defined by the points $\{P_1,P_3\}$. The set of points $\{Q_1,Q_2,Q_3\}$ define a β plane so that β is parallel to α.

We have just constructed a plane parallel to another in a space, for example a Hei space. The same process could be used to construct other orthotropic parallel spaces in general. In spite of each of the considered points being defined by five coordinates, two parallel planes do not have a common point. Here we have begun to sense the need for a definition of parallel planes, but let us go on a bit more.

Let us now consider the set $A = \{P_1,P_2,P_3,P_4\}$, which are distinct points of space (for example, Hei space). If this set A of distinct points does not belong to the same plane, then it defines a Guimel space. Let us consider a point Q_1 so that Q_1 does not satisfy the condition:

$$Q_1 \neq (P_2 - P_1)r + (P_3 - P_1)s + (P_4 - P_1)t + P_1$$

where r, s, and t are real numbers. To put it in another way, Q_1 does not belong to Guimel space defined by A. As we have done above, we can now draw through Q_1 a line parallel to $\{P_1,P_2\}$. Let us take a generic point Q_2 in this parallel line different from Q_1. In this way we can find Q_3 and Q_4. The set of points $B = \{Q_1,Q_2,Q_3,Q_4\}$ will define a Guimel space parallel to the Guimel space defined by A.

Definition:
Two orthotropic spaces of any dimensions are parallel if no point P of the one belongs to the other (no-crossing condition) and if there is an $n + 1$ dimensional space that contains all the points of the two given spaces.

The examples just shown justify the definition.

It is not necessary for the two orthotropic spaces to have the same dimension n.

Definition:

Two orthotropic spaces of dimension m and n where $m > n$ are parallels if: there is no P point common to both and if there is a space of $m + 1$ dimensions that contain all the points of both spaces

Using a straight line and a plane, it is easy to see what happens. Let us consider the ground as a plane of study and take any one point that is in space but not on the ground. Because we are in Guimel space, if we pass a straight line through this point, only one of two things can happen: either the line meets the plane or the line is parallel to the plane. But if we were in Daled space, the straight line might not meet the plane and might not be parallel to it. Indeed, the straight line would then be reverse to the plane.

Definition:

Two orthotropic spaces that do not have a point in common and are not parallel can be called reverse.

Let us study the case of two straight lines. In order to be parallel, they will necessarily have to be on the same plane. That is Euclidian geometry. If they have no point in common, and if there is no plane to contain them, they are reverse.

In our Guimel space, two planes (Bhet spaces) either cross each other or are parallel. For want of dimensions we cannot see reverse planes, but it is easy to give equations defining reverse planes in Daled space.

To proceed with this reasoning, considering a Daled space we can use a point that does not belong to our Guimel space to build another Guimel space parallel to ours. But there is another Guimel space that has no point in common with ours, and it is not possible to determine a Daled space that would contain the two Guimel spaces. If there is a Hei space that can contain all the points of the two Guimel spaces, they will be termed reverse of the first order or simply reverse. We can find Guimel spaces that will not fit in any Hei spaces, but if they can be made to fit into a Vav space, they will be termed reverse of the second order, and so on.

I shall now share a vital little secret with you. These expressions of analytical geometry help us to see and understand hypergeometry. Because we cannot make drawings, we need a guide, a norm to help us reason correctly. This help is extended to us by analytical equations. Studying the equations, we can begin to feel what is happening, and with time, once you get the feel for it, images of the happenings will begin to form in your mind, but they will always be guided by the equations.

In this chapter I have presented the analytical expression of a straight line in a multidimensional space, of planes, and of our own Guimel space.

You have also learned the generalized concept of parallelism and the generalized formula of distance.

Chapter 3
Intersections of Orthotropic Spaces

Axiom:
Whenever there is an intersection of two orthotropic spaces, the intersection will also be an orthotropic space.

Very little is known about orthotropic spaces that cross one another, because the world we live in is very limited as far as dimensions. Once again I suggest always trying an analytical solution as a means to visualize what happens in multidimensional spaces.

When do two geometrical structures cross one another?

The answer is very simple:

This happens when there is at least one point P common to the two structures, and points of each one that do not belong to the other.

In general:
The set of points common to the two structures is called intersection. We may use the symbol $\cap$ from the set theory, which proves itself quite convenient in this case.

Let us begin with two straight lines, two Alef spaces. In order not to be identical, two straight lines cannot contain two common points; they can only have one common point or none. But two straight lines can still be parallel or reversed. This is the moment to understand more clearly what it means to be parallel or reversed, in their different orders.

The intersection of two structures, and particularly of two orthotropic spaces, represents the degree of relationship between the two up to the point of becoming identical or being contained within one another. Let us imagine two orthotropic spaces, E_m of independent m dimensions and E_n of independent n dimensions, with $m > n$. I shall begin the study supposing that E_m and E_n are parallels. What does "E_m parallel to E_n" mean? Firstly it means there is no P point common to E_m and E_n, and secondly there is an orthotropic space $E_{(m+1)}$ that contains E_m and E_n. This demonstrates a certain relationship between the two spaces.

Let us now imagine them reversed. There is now no longer an $(m+1)$ dimensional space that can contain all points of E_m and E_n. As we can see, these two spaces are now somewhat more unaligned. If there is an $E_{(m+2)}$ space that contains all the points E_m and E_n then they will be a reverse of the first order, or simply reverse.

There may be a case in which only an $E_{(m+3)}$ space may contain all the points of the two spaces considered. As can be seen, the relationship between E_m and E_n spaces is diminishing. How far can it diminish? The relationship will cease to exist only when $(m + n + 1)$ dimensional space can contain all points of E_m and E_n spaces. In this case they will be reverses of the highest order, and every point of each of them will use coordinates completely different from the points of the other space, depending on whether you use a system of conveniently located coordinates. In this case there will be no relationship between the two spaces.

Reviewing the case of our two straight lines, what might be the dimension of the space where the reverse of the highest order takes place? The answer is quite simple: $1 + 1 + 1 = 3$. Or, to say it differently, there will be no relationship at all between two straight lines, even in a Guimel space. We are referring to reversed straight lines in the space known to all of us.

To consider the case of two planes, or two Bhet spaces, how many dimensions should there be in the space that accommodates two reversed planes of the highest order? The answer: $2 + 2 + 1 = 5$. This implies that we shall have to be in Hei space in order to witness two reversed planes of the highest order. It is therefore not surprising that we have never seen two reversed planes.

But the contrary shows us results that are not less surprising. The intersection of two orthotropic spaces can occur at any degree, which is also surprising. The two spaces E_m and E_n of the above example can cross one another, according to a point, according to a straight line and so on . . . which means that they can also cross according to an E_0, an Alef, Bhet, Guimel or Daled Space . . . . up to where? Up to $E_{(n-1)}$, because in the next order E_n would be contained entirely within E_m; there would not be any point of E_n that could not be contained within E_m.

Let us go on to some examples that demonstrate all this theory more clearly.

From the logical point of view, it is perfectly natural that two Orthotropic Spaces should cross one another according to E_0, Alef spaces, Bhet spaces, Guimel spaces, Daled spaces, et cetera, but not from the intuitive point of view. Let us look at a few structures that cross one another according to an E_0 space, or just a single point. A plane is expressed by

$$P = (P_2 - P_1)r + (P_3 - P_1)s + P_1$$

where r and s are real numbers.

Let us select an α plane defined by the following points: P_1=(0,0,0,0), P_2=(1,0,0,0), and P_3=(0,1,0,0). Then α plane will be given by:

$$P = (1,0,0,0)r + (0,1,0,0)s$$

Let us now select a β plane defined by the following points: Q_1=(0,0,0,0), Q_2=(0,0,1,0), and Q_3=(0,0,0,1). Then β plane will be given by:

$$Q = (0,0,1,0)u + (0,0,0,1)v$$

As we can easily see, $P_1 \equiv Q_1$. *There is* a point common to the two planes that is easily given by α expression making r and s equal to zero, and by β expression making u and v equal to zero. But the common points cease right there, because any real number r different from zero will give us a point with the first coordinate different from zero, and no point of β plane has the first coordinate different from zero. The same applies to s and the second coordinate that will never differ from zero on β plane. The reasoning applies to the points of β plane varying u and v, which will never be able to belong to α plane. Although surprising, because we have never witnessed, in the Guimel space in which we dwell, the intersection of two planes meeting at a single point does exist and is shown. The only difficulty presents itself in that we must be located in at least Daled space to see such a phenomenon. The point $P_1 \equiv Q_1$ has been located in origin to simplify the analytical expression of the two planes, but the fact is absolutely general and applies to any other point P of the Daled space. I shall leave it to the reader's discretion to repeat the same reasoning for another $P_1 \equiv Q_1$ from Daled space.

This is not a logical proof, and it does not have to be one. I am simply pointing out two planes crossing at a single point, a fact that was absolutely unknown before hypergeometry but that does exist, as has been shown.

Since it would be too trivial, I do not deem it necessary to show two planes crossing one another according to a line, or two Bhet spaces that have an Alef space in common. Any geometry book would show us these commonly known intersections.

Let us now go on to Vav space, where we shall see two Guimel spaces meeting at a single point. To do so, let us take the following points to define Guimel space A:

$$P_1 = (0,0,0,0,0,0)$$
$$P_2 = (1,0,0,0,0,0)$$

$P_3 = (0,1,0,0,0,0)$

and $P_4 = (0,0,1,0,0,0)$.

Let us also take the following points to define Guimel space B:

$Q_1 = (0,0,0,0,0,0)$
$Q_2 = (0,0,0,1,0,0)$
$Q_3 = (0,0,0,0,1,0)$

and $Q_4 = (0,0,0,0,0,1)$.

As we see, $P_1 \equiv Q_1$ is already a point common the two spaces A and B, where space A is expressed by:

$P = (1,0,0,0,0,0)r + (0,1,0,0,0,0)s + (0,0,1,0,0,0)t$

and space B is expressed by:

$Q = (0,0,0,1,0,0)u + (0,0,0,0,1,0)v + (0,0,0,0,0,1)w$

where r, s, t, u, v, and w are all real numbers.

The reasoning used to demonstrate that no point A different from P_1 can belong to B is the same that was used in the former case of α and β planes. Any real number r, for example, different from zero gives us a point with the first coordinate different from zero that will never be in B space, as can be seen scrutinizing the expressions of B space; and the same applies to the parameters s and t, and reciprocally to parameters u, v, and w.

To show two Guimel spaces meeting at an Alef space – a straight line – it would suffice to repeat the above example making $Q_2 \equiv P_2$, and automatically the points $P_1 \equiv Q_1$ and $P_2 \equiv Q_2$ will define an Alef space expressed by:

$$l = (P_2 - P_1)p + P_1 = (1,0,0,0,0,0)q$$

or

$$l = (1,0,0,0,0,0)q$$

where q is a real number.

And the two Guimel spaces will be expressed by:

$P = (1,0,0,0,0,0)r + (0,1,0,0,0,0)s + (0,0,1,0,0,0)t$
$Q = (1,0,0,0,0,0)u + (0,0,0,0,1,0)v + (0,0,0,0,0,1)w$

where r, s, t, u, v, and w are real numbers. As we can see, making $s = t = v = w = 0$, we will obtain the expression of the intersection I where P is equal to Q.

The same way making $P_1 \equiv Q_1$, $P_2 \equiv Q_2$, and $P_3 \equiv Q_3$, we will have two Guimel spaces A and B crossing according to a Bhet space, or according to an α plane given by the points P_1, P_2, and P_3, expressed through:

$$I = (1,0,0,0,0,0)p + (0,1,0,0,0,0)q$$

where p and q are real numbers, and

$$P = (1,0,0,0,0,0)r + (0,1,0,0,0,0)s + (0,0,1,0,0,0)t$$
$$\text{and}$$
$$Q = (1,0,0,0,0,0)u + (0,1,0,0,0,0)v + (0,0,0,0,0,1)w$$

where r, s, t, u, v, and w are real numbers.

These specific points were selected to make sure that the analytical expressions would be simple, but the fact can be shown for any points of Vav space.

To show an example of parallelism and reversed planes, let us select as a plane of reference the plane given by the following points:

$P_1 = (0,0,0,0,0,0)$
$P_2 = (1,0,0,0,0,0)$

$$\text{and} \quad P_3 = (0,1,0,0,0,0)$$

So the reference α plane will be expressed by:

$$P = (1,0,0,0,0,0)r + (0,1,0,0,0,0)s$$

First, I shall show you a plane parallel to α plane given by the following points:

$Q_1 = (0,0,1,0,0,0)$
$Q_2 = (1,0,1,0,0,0)$

$$\text{and} \quad Q_3 = (0,1,1,0,0,0) \text{ expressed by:}$$
$$Q = (1,0,0,0,0,0)u + (0,1,0,0,0,0)v + (0,0,1,0,0,0)$$

There is no intersection because all β points have a third coordinate equal to 1, while all α points have a third coordinate equal to zero, which is different from 1. The Guimel space defined by the following points:

$$M_1 = (0,0,0,0,0,0)$$
$$M_2 = (1,0,0,0,0,0)$$
$$M_3 = (0,1,0,0,0,0)$$
$$M_4 = (0,0,1,0,0,0) \text{ expressed by:}$$
$$M = (1,0,0,0,0,0)m + (0,1,0,0,0,0)p + (0,0,1,0,0,0)q$$

will contain both α and β planes. This Guimel space contains the points P_1, P_2, and P_3 that define α plane and Q_1, Q_2, and Q_3 that define β plane, as can easily be verified. So *there is* a $2 + 1 = 3$ dimensional space, which means that there is a Guimel space that contains both planes, and they do not have any point in common. Therefore, by definition, these two planes are parallel.

But let us unalign β plane a bit more, defining it by the following points:

$$Q_1 = (0,0,1,0,0,0)$$
$$Q_2 = (0,0,1,1,0,0)$$
$$Q_3 = (0,1,1,0,0,0) \text{ expressed by:}$$
$$Q = (0,0,0,1,0,0)u + (0,1,0,0,0,0)v + (0,0,1,0,0,0)$$

No Guimel space that could contain all points of these two planes exists; only a Daled space defined by the following points:

$$M_1 = (0,0,0,0,0,0)$$
$$M_2 = (1,0,0,0,0,0)$$
$$M_3 = (0,1,0,0,0,0)$$
$$M_4 = (0,0,1,0,0,0) \text{ and}$$
$$M_5 = (0,0,0,1,0,0) \text{ expressed by:}$$
$$M = (1,0,0,0,0,0)k + (0,1,0,0,0,0)m + (0,0,1,0,0,0)p$$
$$+ (0,0,0,1,0,0)q$$

where k, m, p, and q are real numbers. That this Daled space can contain the two reversed planes of the first order can easily be verified by observing that it contains all points that define α and β planes.

But I can still show you a β plane defined by the following points:

$$Q_1 = (0,0,1,0,0,0)$$
$$Q_2 = (0,0,1,1,0,0)$$
$$Q_3 = (0,0,1,0,1,0) \text{ expressed by:}$$
$$Q = (0,0,0,1,0,0)u + (0,0,0,0,1,0)v + (0,0,1,0,0,0)$$

Once again, there will be no point common to α and β planes, nor will there be a Guimel space to contain them; not even a Daled space that could contain the two planes. In this case we shall need a Hei space defined by the following points:

$$M_1 = (0,0,0,0,0,0)$$
$$M_2 = (1,0,0,0,0,0)$$
$$M_3 = (0,1,0,0,0,0)$$
$$M_4 = (0,0,1,0,0,0)$$
$$M_5 = (0,0,0,1,0,0) \text{ and}$$
$$M_6 = (0,0,0,0,1,0) \text{ expressed by:}$$
$$M = (1,0,0,0,0,0)k + (0,1,0,0,0,0)m + (0,0,1,0,0,0)n$$
$$+ (0,0,0,1,0,0)p + (0,0,0,0,1,0)q$$

The first orthotropic space to succeed in containing α and β planes. These planes will be a reverse of the second order, the highest order in $2 + 2 + 1 = 5$ dimensions, which is the maximum number of dimensions required to contain two planes reversed in the highest order.

We have just shown through simple examples, using privileged positions, how two orthotropic spaces can meet, be parallel, and be reversed.

It is remarkable that parallelism now appears to us to be a virtual crossing, a relationship between two orthotropic spaces that almost gives us a common point.

Chapter 4
Conjugation of Vectors

Unless explicitly determined otherwise, we should place ourselves in Vav space, as has been shown in the previous chapter. For the considered Vav space to be a vector space, we must define two operations, usually addition and multiplication with valid commutative and distributive laws. Also required for those operations are the elements zero and one that will be the respective neutral elements. Once this is done, we shall consequently be faced with the inverse operations subtraction and division. The set of elements with properties briefly described above define a vector space. Given two vectors $\vec{A} = \{a_1, a_2, ..., a_n\}$ and $\vec{B} = \{b_1, b_2, ..., b_n\}$, let us define the vector sum $\vec{A} + \vec{B}$ as usual; $\vec{A} + \vec{B} = \{a_1 + b_1, a_2 + b_2, ..., a_n + b_n\}$ and similarly the difference of vectors. Let us define the product of a vector by a real number r with $r \cdot \vec{A} = \{r \cdot a_1, r \cdot a_2, ..., r \cdot a_n\}$. The product of two vectors is still missing. In vector calculus there are two kinds of product that will be generalized here.

The scalar product will follow exactly the usual rules, so that $\vec{A} \cdot \vec{B} = (a_1 \cdot b_1 + a_2 \cdot b_2 + ... + a_n \cdot b_n)$. This will give us a real number. Consequently the scalar product of two vectors is a real number.

Let us discuss orthogonality. We define two orthogonal vectors as two crossing vectors whose scalar product is zero.

Now let us remember the concept of vector product (cross product). The vector product is defined for two vectors in Guimel space: $\vec{A}$ and $\vec{B}$. So $\vec{A}$ and $\vec{B}$ belong to the Guimel space and contain three coordinates. Under these circumstances, $(\vec{A} \times \vec{B})$ is defined as the determinant of the matrix, which is obtained by placing the coordinates of $\vec{A}$ in the first row, the coordinates of $\vec{B}$ in the second row, and the unary vectors $\vec{i}, \vec{j}$, and $\vec{k}$ in the third row. This way we can obtain a square matrix whose determinant can be calculated.

The determinant will be a vector sum of $\vec{i}, \vec{j}$, and $\vec{k}$ with their respective coefficients, thus defining a third vector $\vec{C}$ not in the same plane as $(\vec{A} \times \vec{B})$.

Let us write this down:

$$\vec{A} \times \vec{B} = \text{Det} \begin{pmatrix} a_1 & a_2 & a_3 \\ b_1 & b_2 & b_3 \\ \vec{i} & \vec{j} & \vec{k} \end{pmatrix}$$

But cross product is only a particular case of a much more complex vector operation that we shall call conjugation of vectors and express through $@(\vec{A}, \vec{B})$. Note that if $\vec{A}$ and $\vec{B}$ belong to the same line (and therefore to the same Alef space); thus $\vec{A} \times \vec{B} = 0$. This also occurs with the conjugation of vectors, but it is essential to transfer all this to our multidimensional spaces.

Definition:
In a given E_n space, and given $(n-1)$ vectors not belonging to $E_{(n-2)}$ space, and given a Cartesian system of coordinates defined by n unary vectors $\vec{i}, \vec{j}, \vec{k}, \vec{l}, \vec{m}, ..., \vec{n}$, the vector obtained by the determinant of a matrix that has as rows the coordinates of $n-1$ vectors while the last row has n unary vectors, this determinant can be called the conjugation of $(n-1)$ vectors. Mathematically:

$$@(\vec{A}, \vec{B}, \vec{C}, ...) = \text{Det} \begin{pmatrix} a_1 & a_2 & \cdots & a_n \\ b_1 & b_2 & \cdots & b_n \\ \cdots & & & \\ \vec{i} & \vec{j} & \cdots & \vec{n} \end{pmatrix}$$

or

$$\vec{X} = @(\vec{A}, \vec{B}, \vec{C}, ...)$$

Let us now discuss $\vec{X}$. If the set of vectors that we wish to conjugate belongs to any $E_{(n-2)}$ space, then $\vec{X}$ will be zero. The demonstration can be made showing a linear combination of the set of rows so that the determinant is zero. Returning to cross product, if the vectors $\vec{A}$ and $\vec{B}$ belong to any $E_{(3-2)} = E_1$ (Alef space), or if they belong to the same line, we know $\vec{A} \times \vec{B} = 0$, and also $@(\vec{A}, \vec{B}) = 0$.

Personally, I prefer to discard the case of the set of vectors belonging to an $E_{(n-2)}$ space in the definition, but if it does happen without losing generality . . . well, we know the result. In the case of n dimensional space, we must eliminate one of the vectors and test in $E_{(n-1)}$ space the possibility of conjugating the remaining vectors. The conjugation of vectors is presented in matrix form, and the knowledge we have of determinants immediately shows us many properties of conjugation. In the case of n dimensions, where we do not find a no-zero conjugation, we must look for a minor of k order in the matrix of the conjugation, select a convenient system of Cartesian coordinates, and apply the definition to the remaining vectors.

Having defined the conjugation in matrix form, we can immediately say that conjugation is anticommutative:

$$\vec{A} \times \vec{B} = -\{\vec{A} \times \vec{B}\}$$

or in Daled space

$$@(\vec{A},\vec{B},\vec{C}) = -@(\vec{B},\vec{A},\vec{C}) = (\vec{B},\vec{C},\vec{A}) \text{ and so on...}$$

From this we have drawn a very significant conclusion: the conjugation of vectors is not commutative. When we alter the order of vectors, there may be a change in the sign for vector conjugation. Another important conclusion: only the sign is changed, not the magnitude.

What is magnitude in n dimensional space? The answer is a simple generalization of the vector magnitude that we already know.

Definition:
Vector magnitude is the positive square root of the scalar product of the vector by itself.

So far, it is all easy and logical. The conjugation of vectors gives us certain results as we expect them.

Theorem:
Given $\vec{X} = @(\vec{A},\vec{B},\vec{C},...)$ then $\vec{X}$ is orthogonal to all the component vectors. Or $\vec{X}.\vec{A} = \vec{X}.\vec{B} = \vec{X}.\vec{C} = ...=$ZERO.

The demonstration is simple and based on the theory of determinants.

When we were in E_n space (chapter one) and wanted to jump to $E_{(n+1)}$ space, we required a single point outside of E_n, and based on that point we could define $E_{(n+1)}$ space. Given n unary vectors of a Cartesian system of coordinates placed in an E_n space, we can find the $(n+1)$ unary vector of $E_{(n+1)}$ space by simply conjugating the n unary vectors of the system of n coordinates.

Conjugate $\vec{i}$, the unary vector of the axis X, by $\vec{j}$ of axis Y, and we shall obtain $\vec{k}$, the unary vector of axis Z. By conjugating again the unary vectors of the axes X, Y, and Z and by placing $\vec{i}, \vec{j}, \vec{k}, \vec{m}$ in the last row, we shall obtain $\vec{m}$, which is the unary vector of the axis W. Even though we still cannot make a drawing, which in the beginning of this book might have seemed a form of violence, we now have – with all the support of analytical geometry and at complete ease – an axis W simultaneously perpendicular to X, Y, and Z!

As an exercise let us show this conjugation:

$$@(\vec{i}, \vec{j}, \vec{k}) = \mathrm{Det} \begin{pmatrix} 1 & 0 & 0 & 0 \\ 0 & 1 & 0 & 0 \\ 0 & 0 & 1 & 0 \\ \vec{i} & \vec{j} & \vec{k} & \vec{m} \end{pmatrix} = \overline{m}$$

At this point I would like to theorize a bit. Let us consider a straight line, and in it a point P. If we are in the line, in an Alef space, we cannot possibly draw a perpendicular to the given line. Let us go on to Bhet space, where we can now draw one and only one straight line perpendicular to the given line at point P. Now we shall proceed to Guimel space. Here we can draw an infinity of straight lines perpendicular to the given line at point P. Indeed, there will be a plane where any straight line passing through P will be perpendicular to the given line. Going on to Daled space, we shall have a Guimel space perpendicular to the given line in which any line and any plane passing through P will be perpendicular to the given line.

We can clearly see that as we gain more dimensions, the possibilities of orthogonality increase. Spaces containing more dimensions are richer and offer us more orthogonal possibilities. In the previous chapter, we saw that this cannot happen in parallelism because parallelism brings an intrinsic limitation where dimensions are concerned. For two straight lines to be parallel, they have to be on the same plane. In other words, they have to be contained in a Bhet space even if we are in a Vav space.

This concept presents us with an equally significant idea. If we find ourselves in a Vav space (while in reality we are in cosmos!) how can we conjugate only two vectors (a cross product)? The answer is quite simple. Limit yourself in a Guimel space within a Vav space. The two vectors will give you three distinct points in Vav space. These three points define a Bhet space. With this Bhet space and a point of Vav space, you will have a Daled space. In this Daled space you can take three unary vectors, with which you can conjugate the two vectors.

The following question arises: when we consider this fourth point in Vav space, doesn't the cross product, in a way, become dependent on this point? The answer is yes. But this should not appear strange to you once perpendicularism gains possibilities in accordance with the increase of dimensions.

We defined magnitude of a vector as a positive square root of the scalar product of the vector by itself. Once magnitude and scalar product have been defined, we are able to define the cosine of angles between two vectors A and B as:

$$\cos \varphi = \frac{\vec{A} \cdot \vec{B}}{|\vec{A}| \cdot |\vec{B}|}$$

Be aware that A and B belong to any n dimensional space. The definition of angles between two vectors is surprisingly simple. It would seem impossible to measure the angle between two vectors in a multidimensional space. But you can easily reason and see this is not so surprising after all. An angle is a measurement of the arc of a circle, which is a drawing on a plane. Two vectors define a Bhet space, and in the plane we can draw a circle with a unary radius between the two vectors; the measurement of this arc is simply the angle we are looking for.

But it is not all as simple as that. We have the tools. We should use them, apply analytical formulae, examine our problem near the origin to obtain simpler algebraic expressions, make projections (which we shall learn how to do in chapter six), and analyse projection sequences before we draw conclusions. Let us not by any means forget that n dimensional spaces are by far "richer" than the space known to us, and not all affirmations, especially affirmations of uniqueness, are valid with more dimensions.

Chapter 5
A Few Structures

There is no such thing as a hyperline, nor is there a hyperplane, but is there a hypersphere? Yes. We can easily write the equation of a hypersphere in Daled space:

$$x^2 + y^2 + z^2 + w^2 = 1$$

where x, y, z, and w are coordinates of the axes X, Y, Z, and W. But I would prefer to quote the equation of the hypersphere as $x^2 + y^2 + z^2 + w^2 < 1$, which would express a set of points that define the hypervolume of the hypersphere, as was done in chapter one when we defined square. Let us remember that on that occasion, we showed that the equations

$0 < x < l$
$0 < y < l$
$0 < z < l$
$0 < w < l$

define the hypercube, where l is the length of the side of the hypercube. These inequalities express the set of points contained in the space in the same way that our three-dimensional cube does.

All solids can be generalized to multidimensional spaces, including a combination of a variety of them.

For example, the implicit equation of a three-dimensional cone is given by:

$$x^2 + y^2 < z$$

You can imagine a structure that carries conical properties and spherical properties. To make it simpler, let us place ourselves in Daled space and give to our three-dimensional sphere of unitary radius $x^2 + y^2 + z^2 < 1$ the cone property of expanding according to an axis, which in this case would be a W axis. In this way we would be expressing our conic hypersphere through $x^2 + y^2 + z^2 < w$. We shall see that, for each point W, we have a "size" of sphere in our three-dimensional concept.

43

Within the spirit of geometry, we may want to apply what we have learnt so far to these new structures, but first let us repeat that we are not able to build a model of these structures only because they are in a dimension greater than ours. But even this difficulty shall be overcome in the next chapter, where we shall introduce means to present them pictorially in a satisfactory way.

What would hypergeometric structures crossing between themselves be like? And what would hypergeometric structures crossing orthotropic spaces be like? The question does not arouse any degree of difficulty. The intersection of any hypergeometric structure with any orthotropic space decreases the dimensions of the intersection to the maximum dimensions of the intercepting orthotropic space.

For example, the intersection of any structure with a plane, a Bhet space, will bring the intersection to the plane. And since hyperplanes do not exist, you can place yourself on the ground of your backyard to observe, to work, to draw straight lines, or to do whatever you wish on the intercepting plane, including defining a local system of coordinates in order to work more freely. The intersection of a hypergeometric structure with our Guimel space will be a perfectly visible and normal geometric solid. For example, a hypersphere of unitary radius, whose equation we have just shown, crossing with a Guimel space will result in a perfectly normal sphere. Note that a simple sphere in our Guimel space can in reality be an intersection of a hypersphere with our space; in other words, something much more complex than it seems at first sight. Be careful!

Taking one more step, a sphere crossed by any plane gives us a circle that is a plane figure reducing the intersection dimensions to the dimensions of a Bhet space. The same way, upon being crossed by a straight line, an Alef space, the sphere will give a segment of the straight line contained within the Alef space.

I am quite fond of torus. With the liberty that hypergeometry offers, I could show you a torus that expands its major radius according to the W axis and its minor radius according to the V axis. But since the expression of torus even according to the X, Y, and Z axes is not simple, I shall avoid presenting an analytical expression of a hyperdimensional torus.

For simplicity's sake, let us just refer to these structures.

What is a torus? It is a solid generated by a dislocated circle rotating on one axis. We may imagine a sphere dislocated from origin rotating on the V axis. It would also be a hypertorus, but then it would be much more complex.

Now, dear reader, it all depends on our imagination, for the path is open
. . .

Now that we have looked at some hypergeometric structure, we should go on to generalize the concept of curve and surface as we know them.

Definition:
Two sets are isogenic when we can map each element of the one with each element of the other. Mathematically, two sets can be called isogenic when a "one to one" mapping is possible. In Greek, *iso* means "the same", and *genos* means "family", so isogenic should transmit the idea of the two sets belonging to the same family. Note that the two sets in general are not finite, but quite the contrary.

As a first example, with a view to explaining what is being defined as isogeny, I shall present a set of real numbers and the set of points of a straight line. A second example is a set of the complex numbers and the set of points of a plane. It may seem strange, but there is isogeny between the set of points of the open interval (0,1) and the set of all points of the straight line.

In the same way, we can find a function that maps all points of a circle without a borderline with all points of a plane. Therefore, a circle without a borderline and the plane are isogenic.

A plane and the shell of a sphere are also isogenic. We have just seen that in all the given examples there is a common relationship between the pairs of sets: Each pair contains the same number of points $\aleph_0$ or $\aleph_1$, depending on the case. And therein lies the secret of the possible mapping.

Let us consider a curve on a plane. Observe that there is always Isogeny between a curve and a straight line – an Alef Space, just as there is between an open or closed curve in space, especially because we can always cut a closed curve.

Definition:
A curve is a geometric entity isogenic to Alef space.

This definition is absolute, irrelevant of the space in which the curve might be located. Just as hyperlines do not exist, any set in any space isogenic to Alef space is by definition a curve.

Definition:
Surface is a geometric entity isogenic to a Bhet space.

The points of a plane bordered by a closed curve shall be called a plane figure. But a bumpy figure does not fit on a plane, and yet it is isogenic to a Bhet space; consequently, it is a surface.

In order to avoid confusion, consider these geometric entities without a border, because the border is isogenic to Alef and the figure is isogenic to Bhet. After all, any set in any space isogenic to Bhet space is by definition a surface.

Definition:
Solid is a geometric entity isogenic to Guimel space.

We ought to consider a solid a sphere without the shell and without border. A set of points in any space isogenic to a Guimel space is a solid.

Definition:
For sets of points isogenic to spaces of dimensions higher than the three dimensions, we herewith give the name hypergeometric structures, or simply structures.

Observe the sequence.

In the straight line, we have the open interval bordered by two spaces E_0, bordered by two points.

In the plane we have a plane figure bordered by a closed curve isogenic to Alef space.

In Guimel space we have solids bordered by closed surfaces isogenic to Bhet space.

In general, we have hypergeometric structures isogenic to an E_n space bordered by a space isogenic to $E_{(n-1)}$ space.

Note that an orthotropic space isogenic to another space does not cause the latter to become orthotropic. An open interval in a straight line is not orthotropic because it is limited.

Let us now analytically define what we have just shown. In chapter two we saw expressions for Alef, Bhet, and other spaces. In general we can analytically define an orthotropic space by defining a generic point P given by:

$$P = P_0 + \sum_{i=1}^{n} (Pi - P_0)t_i$$

Note that P_0 plus any terms of the sum give us nothing other than the expression of a straight line. Here you see the reason why orthotropic spaces are so akin to straight lines. Note, too, that in truth this expression is equivalent to a system of n equations while the sum contains $i + 1$ terms because we are expressing here an orthotropic space of i dimension in an n dimensional space. To make it clearer, although we have n equations, each one of them only has i parameters (see chapter one, last paragraph).

For a curve to be isogenic to an Alef space, it implies that the curve be described by only one parameter, like Alef space. So each coordinate of a generic point P of the curve should be able to be a function of only one variable m, which can be expressed by

$$P(x_1, x_2, ..., x_n) = P[f_1(m), f_2(m), ..., f_n(m)]$$

where m is a real number contained within an open interval (isogenic to Alef). This is the analytical expression of a curve in an E_n space. To discuss the continuity of the curve is beyond the purpose of this book and would take us to infinitesimal hypergeometric calculus.

In the same way a surface in an E_n space would have points with n coordinates which are functions of two variables m and n, which can be expressed by

$$P(x_1, x_2, ..., x_n) = P[f_1(m, n), f_2(m, n), ..., f_n(m, n)]$$

being that each of these functions would have as a domain a plane figure without its borderline (isogenic to Bhet).

And a solid in E_n space would be defined by three variables m, n, and p, each of them being a real number, while the domain of the function that defines each coordinate will be a solid space without its shell that is isogenic to Guimel space.

We have just shown important hypergeometric structures and their analytical expressions so that they can be useful tools in studies involving hypergeometry.

Chapter 6
Study of Projections

Projections make an extremely useful tool in the study and understanding of hypergeometric structure. Here is a little secret that shows us *how* to understand structures that at first sight are very complex. The projection of a structure is a way to *show* the structure with a smaller number of dimensions. On a sheet of paper (a Bhet space), a plan view of any three-dimensional object fails to transmit all the information about the object, in geometry referred to as solid. Exactly how can we show a three-dimensional solid in Bhet space (the sheet of paper) while it is actually in Guimel space? The answer to this problem can be found in architecture within two well-known approaches: top view and front view. When an architect studies a top view, he gradually forms a three-dimensional view of a house in his mind. The great skill he acquires through time and experience enables the architect to easily visualise the house by simply looking at a blueprint.

In hypergeometry we constantly see structures that cannot possibly be in our Guimel space but, once transformed into projections, can be shown even on a piece of paper! And that is the end (or the beginning?) of our trouble with multidimensional structures. With the exact formula and with the possibility to show any structure on a sheet of paper, the difficulty involving hypergeometry becomes trivial, and making it trivial is exactly the aim of this book.

Let us continue to study the case of the drawing of a house. Let us place X and Y axes on the floor, so that the coordinates of the Z axis indicate the height of each point. What is a top view? A top view is nothing more than the image of a house without the information transmitted by the Z axis. Seen from far above (orthogonally), everything is reduced to the ground and is thus superimposed. Well, there is now only one little problem: the roof of the house obstructs our inner view of the house. This leaves us with only one of two alternatives: either cut the roof of the house – if we really need to see it – and make a separate drawing of it, or cut the house by a plane parallel to the floor, at half the height of the walls. Once again, if necessary, we can make separate drawings of the floor and the roof.

What we really mean to say is that the number of required two-dimensional images does not really matter; some given number of them will transmit an image – with all the necessary details – of the three-dimensional house we wish to view.

In order to make projections, we may use two completely different paths: to suppress a dimension or to cross the solid by a plane, which is a space of smaller dimensions.

Let us begin our examples with a four-dimensional hypersphere of unitary radius with its centre at origin, whose implicit form is

$$x^2 + y^2 + z^2 + w^2 < 1$$

Let us suppress the variable w, making $w = 0$, and we shall obtain

$$x^2 + y^2 + z^2 + < 1$$

which is the known implicit form of a three-dimensional sphere of unitary radius with its centre at origin. Let us observe what happens if we suppress one more variable z, making $z = 0$, and we shall have

$$x^2 + y^2 < 1$$

We now have a circle of unitary radius and centre at origin, and finally making $y = 0$, we shall acquire the open interval (-1,1).

When we are able to draw it, we shall easily see that the projection of the sphere on the plane is a circle like the one obtained, just as the projection of this circle on X axis is simply an open interval (-1,1).

As a rule, it is useful to examine the analytical expression of the structure when we fix the values of some variables – for example zero; but not just zero – giving to the structure the expression of a three-dimensional solid. Let us leave three axes free each time (Guimel space) and repeat the process as many times as necessary, so that we can create an image of the multidimensional structure. The process can be quite irksome in a structure without symmetry in a space of many dimensions, but it is not impossible.

The other process consists of cutting the structure with Bhet and/or Guimel spaces well chosen, making the intersection between them and the structure. Let us now cut the sphere with two planes. Given the sphere $x^2 + y^2 + z^2 < 1$, let us take the plane defined by the points $P_1(0,0,0)$, $P_2(1,0,0)$, and $P_3(0,1,0)$ that is no more than the plane XY. Simple observation leads us to the conclusion that this plane is characterized by $z = 0$. Joining this condition to the expression of the sphere, we shall obtain a circle of unitary radius and with centre at origin. Let us now consider another plane defined by

49

the points $Q_1(0,0,0.5)$, $Q_2(1,0,0.5)$, and $Q_3(0,1,0.5)$; the same way, simple observation shows us that we are dealing with a plane parallel to XY plane crossing Z axis at point $Q_1(0,0,0.5)$. This plane is characterized by the set of points with z coordinate $z = 0.5$. Applying this condition to the expression of the sphere, we shall obtain a circle of radius $\sqrt{0.75}$ and centre at origin. The first circle represents the projection of the sphere XY plane and has been obtained through a crossing that, in turn, was obtained by fixing a coordinate to value 0. The second circle still represents a projection of the part of the sphere that remained above the plane at level $z = 0.5$. The selection of the crossing planes should be based on criteria while the interpretation of the intersection should be done with care. As a rule, the acquired projection making coordinates equal to zero gives us a projection of the whole structure, while the intersection with selected spaces (usually parallels) gives us the spatial evolution of the structure.

Let us go on to another example: $x^2 + y^2 + z^2 < w$. We could call this structure a spherical hypercone. In fact, making any one of the coordinates x, y, or z equal to zero, we shall have a Guimel space with a cone, for example $x^2 + y^2 < w$. Making $w = 0$, we shall obtain only the point $P(0,0,0)$. For any other Guimel space obtained, making $w > 0$, we shall obtain a sphere of radius $\sqrt{w}$. Carefully examining these projections at a given moment, *the impossible image of the spherical hypercone will form in our minds!*

As a rule we can apply the following procedure: Let us imagine any one structure in any one space. Suppose that we seek its projection on XY plane (Bhet space). Let us now take $f(x,y,z,w,\ldots) < 0$ to be the expression that expresses the considered structure. As a habit, I suggest expressing structures through inequalities as equalities take us to the border, and a border has one dimension less than the whole structure. We should issue values to z, w and the others that maintain inequality, and look for the values of x and y that maintain the inequality. This procedure can be made by numerical methods using computers and the set of values $\{x,y\}$ obtained shall be the *shadow* of the structure on XY plane. This process is quite general and can be used to project multidimensional structure to our Guimel space.

I beg your pardon, dear reader, but I simply cannot resist the temptation to show you a hypergeometric meaning of certain functions studied in the twenty-first century whose understanding has remained slightly obscure until today. The example that follows is not only a good presentation of how to apply Daled space, but it also shows that special conditions can reduce the required number of dimensions to describe a given geometric condition. Not only projections decrease the number of dimensions required to visualize a given hypergeometric structure; special conditions do, too, and well beyond our noticing it. On occasion, when we impose special conditions, we are really reducing the number of dimensions of our object of study, and as we are dealing with particular conditions, we shall discover new and unexpected properties.

Hypergeometric Meaning of Analytical Functions

Let us define a function F(z) where z is a complex number as $z = (a + bi)$.
Let us also suppose that F(z) varies on complex plane. So we have a complex function of a complex variable. Let us now use our knowledge of hypergeometry. We are faced with a case that requires at least a Daled space: two dimensions for the domain of the function, one axis for the real part, another axis for the imaginary part of the domain, and two more axes with the same purpose for the range of the function. Consult a good textbook on functions of complex variables, like one written by Ruel V. Churchill, and you will learn that when these functions meet certain conditions, called Cauchy-Riemann conditions, they will be denominated analytical functions. In order to show these conditions, let us suppose that F(z) can be expressed by:

$$F(z) = u(x,y) + v(x,y)i$$

where the functions $u(x,y)$ and $v(x,y)$ are real functions with domain XY plane. Even though it looks as if we have brought the domain of the function to real domain, we still require a Bhet space to define it. I shall put it down in explicit form to clearly show the axes required by the function. Remember that

$$z = a + bi$$

so:

$$F(a + bi) = u(x,y) + v(x,y)i$$

We can now show the Cauchy-Riemann conditions:

$$\frac{\partial u}{\partial x} = \frac{\partial v}{\partial v} \text{ and } \frac{\partial u}{\partial y} = \frac{\partial v}{\partial x}$$

When shown, as above, in the form of partial derivatives, it is quite difficult to understand the Cauchy-Riemann conditions. Churchill teaches us a theorem that establishes the following for analytical functions:

$$\frac{dF(z)}{dz} = \frac{\partial u}{\partial x} + i\frac{\partial v}{\partial x}$$

we can expand:

$$\lim_{\substack{\Delta a \to 0 \\ \Delta b \to 0}} \left(\frac{F(a_0 + \Delta a + (b_0 + \Delta b)i) - F(a_0 + b_0 i)}{\Delta a + \Delta b i} \right) = \lim_{\Delta x \to 0} \left(\frac{\Delta u(x, y_0)}{\Delta x} + i \frac{\Delta v(x, y_0)}{\Delta x} \right)$$

Let us analyse the above expression.

If we wish to express the left branch of equality in space, we must use a Daled space: a Bhet space for ΔF and another Bhet space for Δz. But surprisingly, the right branch of the equation can be mapped on an Alef space: Δu represents the range of the function according to its real axis, while Δv represents the range of the function according to its imaginary axis. But these variations are taken for the same Δx that is *one* of the variables that control $u(x,y)$ and $v(x,y)$. Remember the concept of curves in space and isogeny to Alef space. Strictly speaking, we could use just $u(x,y_0)$ and Δx to calculate the derivative of the real part of the function, and we would then be working on a plane with ordinate $u(x,y_0)$ and x. The same way, we could do the same for $v(x,y_0)$ and the same variable x to calculate the derivative of the imaginary part of the function $F(z)$, which means that each term of the function would be placed on a Bhet space. But $u(x)$, $v(x)$, and x can fit in a Guimel space.

Functions of complex variables require Daled spaces in order to be represented. Their derivatives also require Daled spaces in order to be represented. A certain special kind of functions of complex variables called analytical functions possess a very interesting property: for each point, their derivative can be calculated in a Guimel space, one dimension less than a Daled space. Perceive what a remarkable result this is! What a strong restriction for analytical functions, which may now be considered a very restricted subset of functions of complex variables!

Chapter 7
An Application in the Theory of Relativity

Solely with the purpose of showing that a multidimensional geometry can be useful, let us give a hypergeometrical model of the contraction of space in the theory of relativity. We know that the phenomenon of contraction of space occurs when speeds are near the speed of light. For a crewmember inside a spacecraft of length L in *his* view, the spacecraft will appear smaller than L for the stationary observer. The equation that gives the apparent length *l* of the spaceship to the eyes of the stationary observer is:

$$l = L \sqrt{1 - \frac{V^2}{c^2}}$$

where V is the speed of the spacecraft and c is the speed of light.

What I am about to show you is not a theory of physics but simply a demonstration of the usefulness of hypergeometry.

Consider a classical reference system of axes. We shall add X', that I shall herewith name the relativistic X axis, to axes X, Y, and Z. Let us suppose that the ordinate of X' is a function of the length of the object in XX' plane. Remember that all of Euclid's geometry can be applied on the XX' plane. In the space in which we live, we can only observe the coordinates of X, Y, and Z.

But now we have a Daled space, given that what we observe in the ordinate X is the projection in X of the length of the spacecraft in plane XX'. Let us suppose that the relativistic length is L multiplied by V and divided by c, where V is the speed of the spacecraft and c is the speed of light.

We now have the following: a spacecraft of length L travelling at V speed, whose apparent length is *l*. The coordinates in plane XX' are: *l* on X axis and £ = L(V/c) on X' axis. Working on the XX' plane, it is evident that the length of the spacecraft is L, which is the square root of the sum of the squares of the two ordinates X and X' on the XX' plane (Pythagoras's theorem).

Let us call the ordinate X' relativistic length given by £ = L(V/c). As we can see, the greater the speed of the spacecraft, the bigger its relativistic length. On XX' plane, let the angle between the spacecraft and the axis on which we see its projection in our Guimel space be called η. So

$\sin\eta = (L(V/c))/L$ or $\sin\eta = V/c$. We can call η the relativistic angle of the spaceship because

$$l = L\cos\eta$$

In other words, we see the projection of the spaceship of length L making an angle η with our X axis. Observe that we have a rectangular triangle whose catheti are l and £, and whose hypotenuse is L, with L being the length of the spaceship in the eyes of the crew-member, l being the length of the stationary observer, and £ being the relativistic length of the spaceship. One can easily calculate l in function of L and arrive at the first equation. Note that there is another rectangular triangle similar to the first suggested by $\eta = V/c$ with a hypotenuse equal to c and one cathetus equal to V. And how could I name the other cathetus equal to $\sqrt{c^2 - V^2}$ that by dimensional analysis has dimensions of speed?

This is a book of mathematics, not of physics. Physical phenomena expressed by mathematical formulae use mathematical models to demonstrate the phenomena. If the formula works, even within certain limits, we say that a physical phenomenon can be represented by that mathematical expression. But in reality, they are completely different matters, so much so that the formula only works within certain limits. And humanity is learning increasingly more about this truth. Take, for example, the apple that fell on Newton's head. Any high school student is able to express the speed of the apple, which is $v = \gamma t$, given that the starting speed is zero. But γ is equal to g, so $v = gt$. Again, the equation only works within certain limits. This equation is no more than an algebraic model used to calculate the speed of the apple in function of the time it took to fall. It is a model, just like any other mathematical model, and it is limited . . . Had the apple fallen from space like a meteorite, we could not use the same equation because g would not be constant, since g originates from the gravitational expression

$$F = G\frac{m_1 m_2}{d^2} \text{ where } g = G\frac{m_1}{d^2}$$

where m_1 is the mass of the earth and d is the distance between the earth's centre of gravity and the object's centre of gravity. As we have just seen, g is not constant; g varies with distance d. For the falling apple, the hypothesis of g as a constant is good enough, so the expression that does not suffice for the meteorite is good enough for the apple. In mathematical precision, however, the expression is not good for the apple either, but we are quite happy with it. In the same way, this hypergeometric presentation does not aim to give any physical explanation of space contraction (after all, this is a book of mathematics), but it does introduce a geometric (or rather hypergeometric) model that works and can be used.

Using hypergeometry, we can imagine that, at high speeds almost reaching the speed of light, the length of a moving object, in spite of maintaining its length in the XX' plane while it moves away from the X axis, forms with the X axis at an angle η. In our Guimel space, we see its projection in X axis smaller than its real length. But could it be that, in certain conditions, the spacecraft would dive into the XX' plane, losing its X length and thus turning into photons of light?

As has been said in the beginning of this chapter, this is not a new theory but is just an interpretation of a new geometry that is now possible, and it only serves to demonstrate the applicability of hypergeometry. This model may have more utilities than to simply show a new way to represent the contraction of relativistic space. But at this moment, my intention is no more than to introduce to you a new, possible model.

Finally, you may ask me what is the use of this mathematical puzzle?

This is not a mathematical puzzle.

The real world has many dimensions.

This is a model of a real world.

May I ask you why, in the first moment of the universe, did matter not stay crushed within itself and so remain forever? Because in that moment the universe experienced many dimensions, and it could not be impeded by gravity or any other forces as it escaped through space and time.

And what is going on today?

As space and time dimensions became smaller, other dimensions became important. You cannot see them, and your instruments cannot detect them, but your mind can follow what is happening. Only your mind, as if looking at static pictures based on facts, can realise what happened . . . if you have the right model!

Appendix

Bon-bon

Check Your Understanding

Consider a Daled space. The hypercube given in this Daled space has a starting point $P_1(0,0,0,0)$ and length of side equal to one. In these conditions, answer the following questions:

a) The point $P_1(0,0,0,0)$ belongs to how many faces?

b) What is the farthest point from P_1? What is the distance from P_1 to the farthest point?

c) Give all possible lengths of internal diagonals.

d) How many corners does this hypercube have? List them.

e) How many edges does this hypercube have? List them.

f) Give the parallel, and the reverse edges. Show your work.

g) Give all possible diagonals, internal and external.

h) How many faces does this hypercube have?

i) What faces are parallel? Show your work. List the faces by their corners.

j) Are there any reverse faces? Which ones?

k) Give the intersection of this hypercube and the Bhet space given by:$P_1(0,0,0,0)$, $P_2(0,1,0,0)$, and $P_3(1,1,1,1)$.

l) Give the intersection of this hypercube and the Guimel space given by: $P_1(0,0,0,0)$, $P_2(1,0,0,0)$, $P_3(0,1,0,0)$, and $P_4(0,0,1,0)$.

m) The centre of the previous intersection is $Q(0.5,0.5,0.5,0)$. Show that the distance from Q to an external point (a point not belonging to the hypercube) is zero.

n) Give the intersection of this hypercube and the Guimel space given by: $P_1(0,0,0,0)$, $P_2(1,0,0,0)$, $P_3(0,1,0,0)$, and $P_4(1,1,1,1)$.

o) How many dimensions do the borders of this hypercube have? If the edge is 1 meter, give the measure of the borders $(m, m^2, m^3, m^4, ..., m^n)$.

p) What is the meaning of the sum of the areas of all faces?

www.ingramcontent.com/pod-product-compliance
Lightning Source LLC
Chambersburg PA
CBHW021040180526
45163CB00005B/2205